WITHDRAWN

Looking at Minibeasts

Crabs and Crustaceans

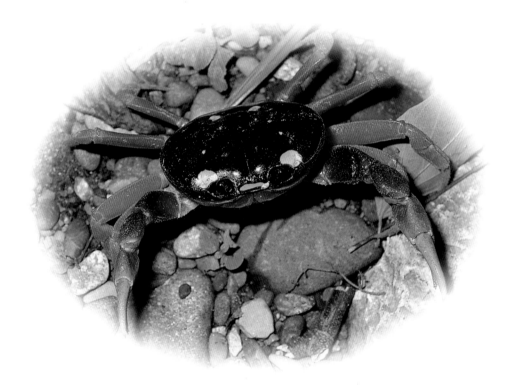

Sally Morgan

Thameside Press

Contents

Words in **bold** are
explained in the
glossary on page 31.

What is a crab?

A crab has a soft body that is protected by a hard shell. It has ten legs that stick out from under its shell. The front two legs are larger than the others, and they have dangerous-looking claws. Crabs are **invertebrates**, which means that they have no **backbone**.

This strange-looking arrowhead crab is found on coral reefs.

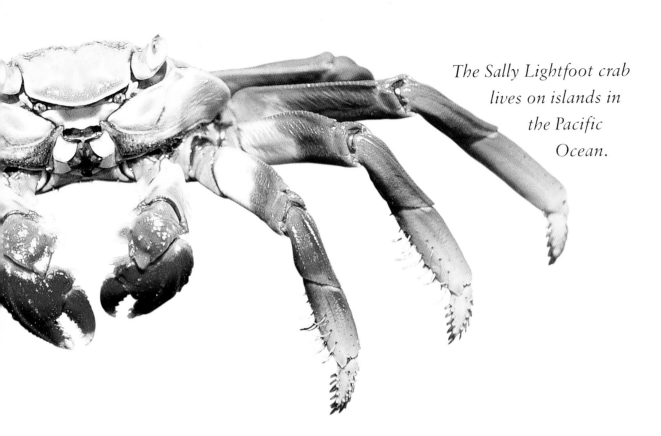

The Sally Lightfoot crab lives on islands in the Pacific Ocean.

Most crabs live in water, but a few can live on land. Spider crabs (see page 10) are the largest crabs. Their legs can be over three feet (1m) long. The smallest crabs are just half an inch (1cm) long.

Some crabs, such as this forest crab, spend most of their time on land.

The crab family

Crabs belong to a group of animals called **crustaceans**. A crustacean is an animal that has a hard outer covering to protect its body. The heaviest crustacean is a lobster that can weigh as much as 45 pounds (20 kilograms).

Lobsters have large, powerful claws.

During the day wood lice hide in damp places under stones and logs. They come out at night to eat.

The wood louse lives on land. It is a small, flat animal. Its body is protected by a hard outer covering, and it has seven pairs of legs.

Water fleas and copepods are **microscopic** crustaceans that live in pond water.

Water fleas have a see-through shell. You can see their insides.

Shell shapes

A crab has a hard shell to protect its soft body.

A crab's shell can be many different shapes,

colors, and sizes. Some shells have a jagged

edge and others have a wavy edge.

Some shells are bumpy and some

are smooth.

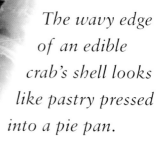

The wavy edge of an edible crab's shell looks like pastry pressed into a pie pan.

This is a shore crab. It has a shell with five jagged points on each side.

A decorator crab covers its shell with bits of coral so that it cannot be seen by **predators**. Some crabs have **sea anemones** living on their shells. The sea anemones help protect these crabs from predators.

These decorator crabs have decorated their shells with different colored corals.

Ten legs

Crabs and lobsters have ten legs. Their legs have **joints** so that they can bend. The legs of a crab stick out to the side. It cannot move forward, so it has to walk sideways.

A spider crab has very long legs. It is one of the largest crabs on the ocean floor.

Most crabs walk around on the sea bed, but they can swim too.

This ghost crab scurries sideways along a beach on the tips of its legs.

   Prawns and shrimps use their legs like paddles to push them through the water. Water fleas and barnacles use their legs to collect food from the water.

Fairy shrimps have feathery "legs" that trap tiny pieces of food floating in the water.

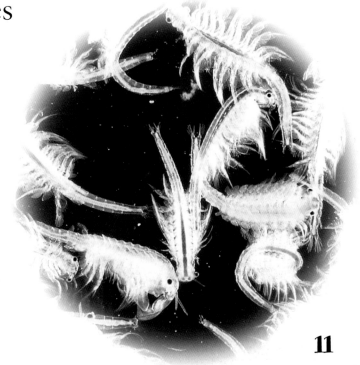

Hiding places

A crab's flat body means that it can squeeze under rocks. It can hide out of sight and then jump out on small fish that swim past.

When a lobster is scared, it shoots backwards into a hole in the rocks and blocks the hole with its claws.

This crab is just the right shape to squeeze into this narrow space.

A pile of sand builds up as a crab digs a burrow using its claws.

Some crabs make small burrows in the sand. When the tide is out, they hide in their burrows. They come out to eat when the tide comes back in.

Few animals would try to pull this squat lobster out of its hole.

Useful claws

A crab has a big claw on each front leg. One of the claws is usually larger than the other. The larger claw is used to crush the shells of mussels and clams, which a crab likes to eat. The smaller claw is used to pick up and handle food.

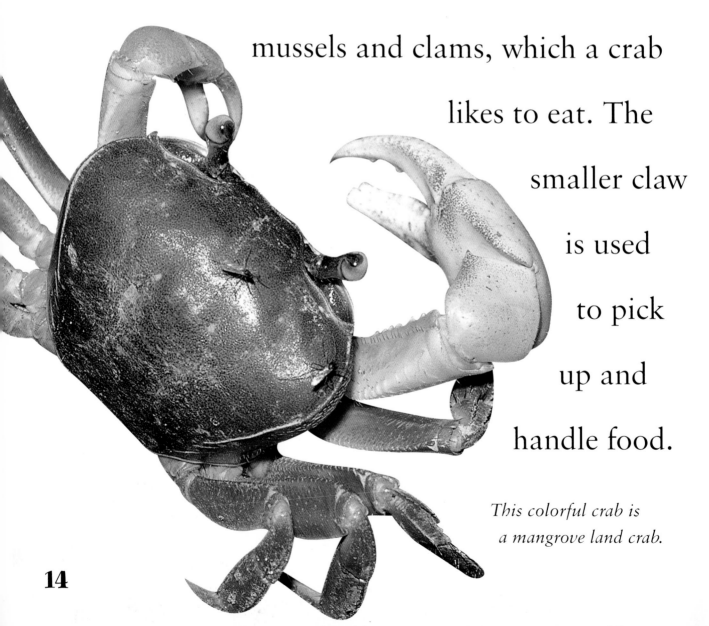

This colorful crab is a mangrove land crab.

When a crab is threatened by another animal it waves its large claws in the air.

Mantis shrimps use their claws to make ear-piercing sounds to scare their enemies.

A crab also uses the larger claw to fight other crabs. Some male crabs use their larger claw to attract female crabs.

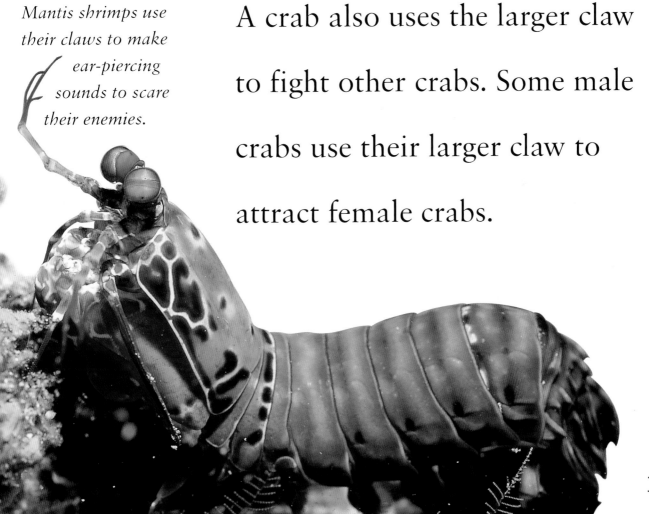

What do crustaceans eat?

Some crabs are **carnivores**. They feed on shellfish on the sea bed such as mussels and clams. They also catch small fish and starfish.

Wood lice are found in piles of rotting logs where they feed on wood, dead leaves, and fungi.

Many crabs are **scavengers**, which means that they feed on dead animals that they find lying on the sea bed. They use their claws to rip the food to pieces.

A crab rips a starfish to pieces with its claws. Then it uses its mouthparts to push the pieces into its mouth.

Seeing and smelling

Crustaceans use their **antennae**, or feelers, to smell. They need a good sense of smell to find food. Shrimps and prawns have very long antennae.

The spiny lobster uses its long, sharp antennae as weapons.

The long white antennae of these shrimps are longer than the animal's body.

Crabs have a pair of eyes on stalks.

They can cover themselves with sand

and still see what

is going on

around them.

The fiddler crab has eyes on long stalks, which means it can see over its large claw.

Prawns and shrimps

Prawns and shrimps have long, round bodies with flattened sides. This makes it easier for them to squeeze into holes in rocks. Tiny cleaner shrimps help other sea creatures by removing **parasites** from their bodies.

This tiny cleaner shrimp is hiding in the tentacles of a sea anemone.

Many shrimps have see-through shells. But some shrimps' shells are the same color as the rocks and coral where they live, to help them hide.

The red spots of this painted shrimp help it hide from predators.

The common prawn has a pale brown shell that is almost see-through. This makes it difficult to spot on a rock.

Life cycles

A female crab lays her eggs and carries them around until the eggs hatch into tiny **larvae**. Larvae and other microscopic animals and plants that live in the sea are called **plankton**. Whales and fish feed on plankton. Crab larvae look very different from adult crabs.

Plankton are so small that they can only be seen under a microscope.

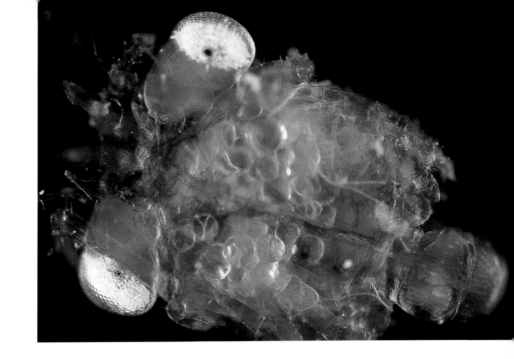

This crab larva looks nothing like its parent right now, but it will grow larger and change shape.

The larvae drift through the water, gradually changing shape until they become small crabs.

A female wood louse keeps her eggs safe by carrying them around with her until they are ready to hatch.

These tiny wood lice have just hatched. They take about two years to grow to full size.

Growing a new shell

As a crab's body grows, it needs a larger shell.

A crab cannot live without its shell. So, before a crab can lose its old shell, it grows a new one under it. Then the crab bursts out of the old shell.

This shore crab (above) is ready to shed its old shell.

Once it has shed its old shell, the crab is much bigger.

This large edible crab is probably several years old. The largest crabs are usually the oldest crabs.

The new shell is soft at first,

and the crab's body swells up in size before the

new shell hardens. A crab will shed its old shell

and grow a new one many times during its life.

Hermit crabs

A hermit crab is unusual because its body is not protected by a shell. A hermit crab has to find an empty shell and squeeze its body into it. The shell protects it from predators.

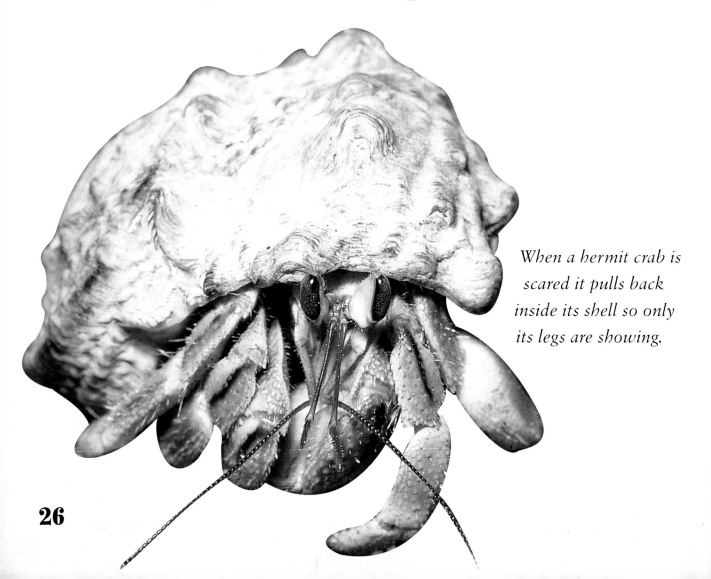

When a hermit crab is scared it pulls back inside its shell so only its legs are showing.

The robber crab, which is the largest land crab, can climb trees. It is a hermit crab, but it does not live in a shell.

When a hermit crab outgrows its shell, it has to find a new, larger one. It looks around until it finds another, bigger shell to make its home.

Hermit crabs have to drag their shell around with them when they hunt for food.

Watching minibeasts

When you fish for crabs, keep very still. Any movement will scare them.

If you walk along a sandy beach you may see small burrows where crabs live. Crabs hide in their burrows while the tide is out and come out to eat when the tide comes back in. Crabs can be found in tide pools too, where they hide under rocks. You can fish for crabs in tide pools by dangling a piece of fish tied to some string in the water. If you sit still, you may see a crab come out to eat the fish.

You can buy cooked crabs to take a closer look at one. Open up the shell and look inside. See how the legs are joined together and how the claw works.

There are many tiny crustaceans living in pond water. In the summer, fill a clear plastic bottle from a pond. Using a magnifying glass, look carefully at the water. You should be able to see small animals moving around in the water. These are water fleas and copepods.

Wood lice are land crustaceans. During the day they hide under logs, where they feed on the rotting wood. Lift up the logs carefully to see how many wood lice you can spot. Make sure you put the logs back where you found them.

Don't forget to put the pond water back when you are finished.

A pill bug is a type of wood louse that rolls up into a ball when it is scared.

Minibeast sizes

Crabs and other crustaceans are many different sizes. The pictures in this book do not show them at their actual size. Below you can see how big some of them are in real life.

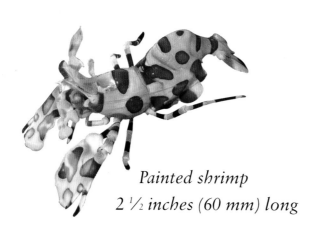

Painted shrimp
2 ½ inches (60 mm) long

Common prawn
2 inches (50 mm) long

Water flea
¹⁄₁₆ inch
(2 mm) long

Wood louse
½ inch (10 mm) long

Hermit crab
1 ¼ (30 mm) inches across

Glossary

antennae Feelers on a crustacean's head.

backbone The strong bones down an animal's back.

carnivore An animal that kills and eats other animals.

crustacean An animal with a hard outer covering to protect its body.

invertebrate An animal that does not have a backbone.

joints Places where two parts slot together so they can move.

larva A young crustacean. It looks very different to an adult.

microscopic Too small to be seen by the eye.

parasite An animal that lives on or in another animal and harms it.

plankton Microscopic animals and plants that float near the surface of oceans.

predator An animal that feeds on other animals.

scavenger An animal that feeds on the remains of dead animals and plants.

sea anemone A tube-shaped animal with stinging tentacles that lives in sea water.

Index

U.S. publication copyright © 2001 Thameside Press

International copyright reserved in all countries.
No part of this book may be reproduced in any
form without written permission from the publisher.

Distributed in the United States by
Smart Apple Media
1980 Lookout Drive
North Mankato, MN 56003

Text by Sally Morgan
Illustrations by Woody

Editors: Claire Edwards, Sue Barraclough
Designer: John Jamieson
Picture researcher: Sally Morgan
Educational consultant: Emma Harvey

ISBN: 1-930643-11-X

Printed in Hong Kong

9 8 7 6 5 4 3 2 1

Library of Congress Cataloging-in-Publication Data

Morgan, Sally.
  Crabs and crustaceans / Sally Morgan.
    p. cm. -- (Looking at minibeasts)
    ISBN 1-930643-11-X
    1. Crabs--Juvenile literature. 2. Crustacea--Juvenile
literature. [1. Crabs. 2. Crustaceans.] I. Title.

  QL444.M33 M67 2001
  595.3'86--dc21

                                    2001023422

Picture acknowledgments:
Jeff Collett/Ecoscene: 15b. Jeff Geoman/NHPA: 27t. Chinch
Gryniewicz/Ecoscene: 13t. Elgar Hay/Ecoscene: 20. Wayne
Lawler/Ecoscene: 17t, 19b. John Liddiard/Ecoscene: 12, 13b,
18, 25. Papilio: front & back cover tl & tcl, 2, 4, 5t, 6, 7t,
7b, 8, 9t, 10, 11t, 11b, 17b, 22, 27b, 30bl, 30br. K.G. Preston-
Mafham/Premaphotos: front cover tr, cl, cr & c, 1, 5b, 14, 15t,
26, 30cr, 30bc. R.A. Preston-Mafham/Premaphotos: 3t, 21b,
24c, 24b. Kjell Sanders/Ecoscene: front & back cover tcr, 3b,
9b, 16, 19t, 21t, 23t, 30cl. Robin Williams/Ecoscene: 23b.